THE YELLOW PERIL

REVILO P. OLIVER

1983
LIBERTY BELL PUBLICATIONS

Reprinted 2005
by
Liberty Bell Publications
PO Box 890
York, SC 29745
www.libertybellpublications.com
ISBN: 1-59364-026-9

ISBN: 0-942094-11-5

LIBERTY BELL PUBLIC.` TIONS
Box 21, Reedy, W.Va. 25270 USA

Printed in the United States of America

The Yellow Peril

by
Revilo P. Oliver

The report from London on Japanese industrial superiority in the January *Liberty Bell* asked, in effect, whether the nations of Europe and North America which are still largely White could do anything about it. That, of course, was the wrong question. The crucial consideration is what, if anything, the Jews will order their Aryan serfs to do about it.

Alert observers in this country have long noted the ominous potential of Japanese industry. In the *New Libertarian* (February-April 1982) the well-known "revisionist" historian, Professor James J. Martin, boldly asked the obvious question that is unthinkable to well-trained Americans: Will their government eventually promote another war against Japan to destroy her industrial superiority? He decided that it would not.

On a quite different level, the periodical *Plain Truth*, published by one of the richest of our holy rabble-rousers to stimulate his business, carried in the issue for February 1983 an article, "Will Century 21 Be the Japanese Century?" It begins by noting that American jewellers assure their customers of the superior quality of watches "totally made in Japan," and that Japanese railroads are the best in the world, with an implied contrast to the railways of which the United States was proud before governmental sabotage began to reduce them to junk. And the article states, as did the British commentator, the basic fact: the average worker in Japan, no matter how menial or banausic his task, "has a sense of responsibility to his job, his employer, *and his country.*" (My italics.) The article is, in fact, a good one until we come to the last paragraphs, where we find the old ballyhoo about "Bible prophecy" and what Yahweh will do for everyone (including Australian Aborigines and African Pygmies!) if only we appease him in the ways known to holy men.

The press in this country has occasionally carried news of Japanese progress. Notably, the *Wall Street Journal* carried a long series of news reports and articles in 1982. There was even a hopeful prediction (13 July) by Peter Drucker that Japanese

3

society would soon be afflicted by the pernicious anaemia that has prostrated us. Several of the factual articles discussed the American government's demands that Japan expand her military forces, ostensibly to counteract the "growing Soviet military presence in the Pacific"—in the territories and on the islands that the American boobs gave the Soviets in 1945. Unmentioned, of course, was the tacit hope that the economic burden of supporting armed forces on a large scale would hamper Japanese industry. It was reported (22 November) that one of the new Japanese warships equipped to fire guided missiles, the *Asakaze*, cost $110,000,000. It was not noted that if the vessel had been built in the United States, it would have cost at least five times as much.

According to some estimates (26 November), Japan, which now produces excellent aircraft, including fighters, has already become the seventh largest manufacturer of military equipment in the world, and will capture the world's markets, if she starts exporting in earnest. Buried in one article (7 June) was a really sensational datum: radar equipment for our F4 Phantom fighter planes made in Japan is three and one-half times more reliable than the same equipment made in the United States. And, what is more, "Japanese versions of U.S. missiles are notably more accurate, thanks to much higher standards of quality control and maintenance."[1]

The average American reader of such items probably lit a cigar and relaxed, speculating about the females he might find at the "happy hour" in his favorite barroom that evening or the probable performance of highly paid entertainers in the next football game. And if he read one item (19 November), he was not surprised by the trite news that Japan can manufacture

1. Articles about Japanese industry appeared in *Business Week,* 14 March 1983, and although their tenor was to give Americans as much reassurance as possible, one article admitted that the Japanese are excelling us not only in accurate manufacture but also in technological research, with the result that this country may soon be in the "unpleasant position of having to rely on Japan for critical military technology." The articles particularly consider the probable effects of American pressure on Japan greatly to increase her own military establishment so that she will be able more adequately to defend herself against the Soviet, and it is only typical of our journalism that there is never a hint that the United States, under the rule of its traitors and fools, deliberately installed the Soviets in the territories from which they now threaten Japan, and that a determination to make the Communists supreme in Asia was one of the purposes for which we waged war against Japan in 1941-1945.

circuits on semi-conductors and sell them "at half of American prices and still make a profit." But if he read consciously to the last paragraph of that article, he may have forgotten to reach for his cigar for several minutes. That paragraph quoted a Japanese official as saying, "The Japanese...can manufacture a product of uniformity and superior quality because the Japanese are *a race of comparatively pure blood, not a mongrelized race as in the United States.*" (My italics.)

That statement of a simple and obvious fact naturally provoked hysteria. Subsequent issues of the *Journal* carried indignant letters about that "shocking comment" and, of course, screeching about "bigotry" and "racism." It was not clear whether this standardized slop came from Jews or high-minded nitwits; it probably came from both. There was, naturally, agreement that Japan must become as righteous and diseased as the United States, rotted with hordes of mongrels, enemy aliens, and black savages, so that Japan can enjoy the blessings of a confiscatory taxation to nourish parasites and speed up their breeding, and enjoy the rapes, "muggings," robberies, and murders which are becoming merely commonplace in a nation in which do-gooders have, in recent years, got the crime rate up to an increase of 12% per annum, and may succeed in boosting it to an annual increase of 18% this year.

Since the Japanese are not humanitarian imbeciles, they will merely chuckle at the raving of the righteous barbarians. But what will the Jews do?

The standard Jewish technique for obtaining possession of the whole world that their god gave them is to induce in other races the mongrelization that will debilitate their victims and render them helpless. Will the Jews tolerate Japan's industrial superiority when it impairs the usefulness of their colonies in Europe and North America, on which the Jews now particularly rely to finance their international terrorism and their war against the Semitic races of the Middle East? That, we may assume, will depend on whether or not the Jews now control Japan as they do the United States, Canada, Britain, and the rest of Europe.

THE DIASPORA

A Jewish hoax that has been quite effective in keeping Christian minds muddled for centuries is the silly story of a "diaspora" caused by the Roman capture of Jerusalem in A.D. 70 and temporary suppression of the Jewish rebellion in

Palestine. That supposedly caused the poor, persecuted Jews to spread to other lands. As a matter of fact, of course, the capital of Jewry in A.D. 70 was in Babylon, outside the Roman Empire; vast hordes of Jews were eating on Egypt, where they were even given special privileges by the stupid Romans; and, as the Jews themselves boasted, they had long before planted their parasitic colonies in every region of the world in which they could conveniently bamboozle and exploit the natives. Their real diaspora had taken place centuries before A.D. 70.

Jews, tentacles of their international race, reached China in the first century B.C., according to their own traditions, and remain there as a force of which we cannot calculate the power. The Jews in China, with their race's peculiar ability to assimilate other races physically while retaining their own racial mentality, are indistinguishable from Mongolians, at least to Western eyes, and their congeners in the West would have us believe that they have "disappeared." But a competent British journalist, Graham Earnshaw, visited Kaifeng, a provincial town in inland China, in which, as he reported in the *Daily Telegraph* (London) on 1 June 1982, the "last synagogue...collapsed in the 1860s," and the Jews no longer circumcize their male offspring to propitiate their bloody god and do not abstain from pork, the flesh of the animal that was probably their totem when Yahweh issued his dietary ordinance. Mr. Earnshaw interviewed a Jew who bore the Chinese name of Shi and told him, "In every way, we are just like the Chinese around us. We look the same, we eat and dress the same, but *I still consider myself to be Jewish.* When I have to fill out the forms on which I have to state my race, I put 'Jew.' " (My italics.) What is more, Shi, although he had never been outside China, had tattooed on his arm a number exactly like the numbers that the diabolical Germans supposedly tattooed on the arms of Jews who now roll up their sleeves and explain that the bungling Germans somehow forgot to shove them into the fabulous gas-chambers in which so many millions of God's Children were exterminated. Shi claimed he had tattooed on his own arm a memorandum of a date he wanted to remember but got the date wrong.

Mr. Earnshaw also noted that there were seven Chinese family names which were regarded as showing Jewish ancestry, at least in Kaifeng, and that "one of them curiously enough is *Jin*, the Chinese word for 'gold,' which figures in so many Jewish names elsewhere, such as Goldstein and Goldberg."

I do not know whether the *Daily Telegraph* was slapped for

6

printing its correspondent's highly significant report from Kaifeng. And if the author of the report from London in *The Liberty Bell* reads the *Daily Telegraph*, he either did not see the possible implications of that article or forbore to mention them.

From China, the wandering Jews—wandering with a purpose —must have moved on to Japan. When they infiltrated Japan, Jewish sources do not tell us, so far as I know, but I certainly do not pretend to a thorough or even extensive knowledge of those sources. In his ably-written and erudite treatise, *The Lost Tribes, a Myth* (Duke University Press, 1930; reprinted, New York, Ktav, 1974), Professor Allen Howard Godbey informed us (pp. 423 sq.) that

> Judaism certainly reached Japan. The extent of its spread and influence is still undetermined....The concrete facts are that in the province of Yamato [= ancient Washū, in the modern prefecture of Kyōto, which surrounds the city of Kyōto, the former capital of Japan and still its third largest city, southwest of Tokyo] there are two ancient villages, Goshen and Menashe (Manasseh). For these names there is no Japanese etymology. The legend is that in the third century of our era a strange people of about one hundred silk raisers appeared. In the census of the year 471, this people numbered eighteen thousand six hundred and seventy and were highly esteemed in the province. A temple known as the "Tent of David" still stands where they first settled. Figures of a lion and a unicorn standing at the entrance are called "Buddha's dogs" by the Japanese....A folk-legend still current says that the founder of the sect, when a child, was found in a little chest floating upon the water. The people today call themselves Chada, "The Beloved." This is traditionally the meaning of "David." But it may reflect "Chasid" [= a member of the Jewish sect of Chasidim, ancient terrorists whom the Romans called Sicarii from their favorite method of murdering civilized men; they are commonly called Zealots, from the Greek word used to designate the terrorists by Josephus and in the "New Testament"].
>
> In the city of Usumasa, on a site belonging to one of the oldest Chada families, is a well some fifteen hundred years old. Upon the stone curbing the word "Israel" is engraved.... The Chada came by way of Korea, where they had an academy in Piang Yang. Its name was Ypulan, in Chinese hieroglyphs [*sic!*]. Professor Anasaki, of the University of

7

Tokyo, considers it the phonetic equivalent of "Ephraim" [!].

I am willing to believe that there is some historical basis for this account.[2] It is quite likely that a passel of Jews penetrated Japan at some early date and acquired control of the highly profitable silk-trade. It is a little astonishing, however, that they should have multiplied in about two centuries to the number of 18,670. It is true that when an advance guard of Jews have fixed their mandibles in a native population, their compatriots swarm in to help in the exploitation and share the profits, but it is hard to suppose that they poured into Japan at a rate which, given the early date and the remoteness of Japan, would be comparable to the way in which they swarmed into the United States to eat on the stupid Americans in the latter part of the Nineteenth Century.

The figure of 18,670 given by the supposed census of 471 can be explained by either (or both) of two techniques that are commonly employed by invading Jews. Male Jews marry wealthy native females, and Jewesses marry wealthy or influential native males, and both sexes use the spouses whom they secretly despise to further the purposes of Yahweh's Holy Race and also engender half-breed children who will be trained to carry on the righteous work under the supervision of pure-blooded Children of the Lord.[3] The Jews also attack their

2. Professor Godbey's footnote shows that his principal source of information was a work by a certain Dr. J. Kreppel, *Juden und Judenthum von Heute* (Vienna, 1926), which I have been unable to procure. Kreppel was, no doubt, a Jew, and since I cannot check his documentation, I cannot guarantee that he did not perpetrate a typical Jewish hoax. I particularly wish I could verify the report of a census in 471, but I must leave that task to someone who can read literary Japanese with ease and has access to the chronicles published in that language. The reference to a "lion and a unicorn" is troublesome: it reminds one that the presence of those two animals as supporters of the British royal escutcheon (which dates from 1707!) was used, during the "British Israelite" craze, as "proof" that the Kings of Britain were descendants of a Jewish bandit named David. The second paragraph in the quotation from Godbey presumably rests on the authority of Professor Anasaki, whom Rabbi Jacob S. Raisin, in his *Gentile Reaction to Jewish Ideals* (New York, Philosophical Library, 1953), p. 422, identifies as "the chief proponent of the Japan-Israel theory," which I shall mention below. The value of Anasaki's evidence is extremely problematical.

3. I wish that the Jewish ban on genetic research had not prevented

selected victims with proselytism, infecting and paralyzing the natives with superstitions cunningly adjusted to their gullibility. The most conspicuous use of that technique in our time is the Bolshevik (Communist) cult, an old Jewish trap baited for modern taste by making it seem irreligious. When the Jews invade a nation, their usual technique is to induce the natives to worship a Jew god and venerate that god's righteous Master Race, with much yammering about the "love" and "brotherhood" the Chosen People are eager to bestow on their destined serfs, and ideally the Jewish "ideals" and deals should make the befuddled proselytes imagine that they can become Jews by being "converted" and submitting to their masters' barbarous regimen. Thus, when conditions are suitable, the Jews spread undisguised Judaism and even admit obedient dogs to their synagogues while privately chuckling over the stupidity of the *goyim*. But the example of Communism, like the archaeological evidence from the excavation of ancient synagogues at Dura Europos and elsewhere, and the Jews' use of the Thracian god Sabazius and the Egyptian Osiris as stalking horses on occasion, should remind us of their great versatility and the ingenuity with which they adapt their bait to the animals they wish to trap. It follows, therefore, that while it is virtually certain that a band of Jewish immigrants would not only use their religion and "righteousness" as a cover for their own activities, but would also delude the Japanese populace with superstition and occult hocus-pocus, we cannot determine *a priori* precisely what form of religion they would induce as most effective in exploiting the weaknesses of the native race.

If we accept the figure of 18,670 for the year 471, we can imagine, in the absence of valid data, that the total includes a nucleus of Jews, a lower caste of half-Jews (presumably offspring of male Jews by native women), and a pack of befuddled Japanese proselytes who suppose that they have been admitted to the privileges the Jew god bestows on his Chosen Race. On that assumption, the figure becomes quite plausible, even modest.

The 18,670 must have left a numerous progeny. What

---

verification or refutation of the alarming claim by Dr. Nossig that even the slightest taint of Jewish blood will pervert the brain cells of our and other races and make the unfortunate descendants susceptible to Jewish manipulation for "many generations." See my *Enemy of Our Enemies* (Liberty Bell Publications, 1981), p. 27, n. 30, for a fuller reference.

became of them? What happened to them during the twelve centuries before Japan came into contact with our race and civilization? So far as I know, the Japanese annals make no mention of them, and if that is so, the Jews and their Judaism must have gone underground or dwindled to insignificance.

As for the Japanese proselytes, we may conjecture that, for one reason or another, many of them did not transmit their infatuation to their descendants and that, in the absence of an effective Jewish control, the cult disappeared in a few generations, except, perhaps, for a few small coteries who, like their counterparts among us, practiced an alien superstition because it was exotic. Between the Fifth and the Seventeenth Centuries of our era, the history of Japan includes many periods of internal turmoil and prolonged civil war, and it is entirely possible that the enclave of Jews in the nation suffered drastic losses of wealth and life, much as large enclaves of Jews in China are said to have been diminished by that nation's internal strife and, perhaps most of all, the Mongol invasion and conquest. The surviving Jews in Japan may have found it expedient to disappear and, with the versipellous talent of their race, become Japanese Marranos, outwardly resembling the natives but secretly aware of the divine ichor in their veins and their enormous racial superiority. The question before us is whether they were sufficiently numerous and adroit to have attained some measure of control over that nation and the formidable racial energies of the Japanese. Although the question is, for want of evidence, insoluble, we may reasonably hope they were not.

Godbey's account implies that although the Chadas, presumably Jews or part-Jews, survived to our time, they are few, an inconsequential survival from the past, comparable, perhaps, to the Jews in Kaifeng. The example of the Jews in China warns us that we cannot rely on the physiognomic and physiological indications of race when dealing with Jews, but, with that proviso, we may observe that there is no evidence of a Jewish element in the native Japanese today. We should notice, however, one effort to provide such evidence.

In the latter part of the Nineteenth Century, Norman McLeod, a pious Scot whose mind had been filled with Judeo-Christian myths, visited Japan and produced his *Epitome of the Ancient History of Japan* (3d edition, Tokyo, 1879), in which he adduced various parallels of custom and belief to prove that the Lost Ten Tribes supposedly abducted by the

Assyrians in 720 B.C. had made a beeline for Japan and there set themselves up as the priests of Shintō ('the divine way'), the native Japanese religion. McLeod added a sheaf of drawings showing, according to his specifications, the rafts on which the Chosen People reached the Nipponian islands and even the order in which the Ten Tribes marched on their way to their new Promised Land. He did, however, present some evidence that can be taken seriously: pictures of contemporary Japanese, some of them with quite prominent noses, whose features he identified as distinctively Hebraic. The value of this evidence is very slight. The pictures, granting the accuracy of the artist who drew them, are not really cogent, and while some of the subjects may be Chadas, it is only too likely that McLeod, his mind buzzing with Jewish fictions and eager to obtain confirmation of them, was misled by the Manchu strain that appears in some Japanese or even by the vestiges of Caucasian (conceivably Aryan) ancestry that are found in a small minority of Japanese and of which the genetic origin can only be conjectured.[4]

4. The Manchus are, of course, a Mongolian (Mongolid) people, but probably with some Turanid admixture, and characteristically have relatively aquiline noses. Japanese anthropologists recognize six distinct ethnic groups in the population (exclusive of Ainu and mongrels) and generally admit that at least two of these show very distinct traces of Europid ancestry, which seem most pronounced in the aristocracy, to which most of the tall Japanese belong. The sources of these admixtures cannot now be identified, but all of the Japanese are predominantly Mongolian (Mongolid), and contemporary Japanese are therefore correct in describing themselves as a comparatively pure race. If our ancestors had had the intelligence rigorously to exclude from our country all immigrants who were not Englishmen, Scots, Germans, Scandinavians, or Nordics from other parts of Europe, we would be today a comparatively pure race, although we would show physical variations comparable to those found among the Japanese and could still distinguish between ethnic strains, noting differences that are now obscured by the great contrast between Aryans and the rest of a population that has been formed by making the United States a dumping ground for all of the world's anthropoid refuse.

On the physical variations found in the subraces of Mongolids, see the fundamental treatise by Dr. John R. Baker, *Race* (Oxford University Press, 1974; reprinted, Athens, Georgia, Foundation for Human Understanding, 1981), pp. 537-539. Baker himself ascertained, from a study of Eighteenth-Century portraits, that women with distinctively Manchu features were considered the paragons of female beauty by the Japanese aristocracy of that time, and the probable result was a kind of selective breeding. Although the Japanese are almost purely Mongolian, some of them inherited White genes. It is a well-known characteristic of the Mongolians that they lack the glands in the armpits and crotch that in

Although McLeod's fantasies are without historical or ethno-
logical value, they have acquired a noteworthy political signifi-
cance. The Jews' myth about "Lost Tribes" spawned the "British
Israel" nonsense, which was so effective in softening English
brains in the time of Disraeli and the massive Jewish contamination of the British upper classes. An adaptation of that hoax is
now being used in an attack on Japan.

"JAPANESE ISRAEL"

The Hungarian writer, Itsván Bakony, in a small booklet
entitled *Jewish Fifth Column in Japan,*[5] believes that if the
tradition about the Jews in Japan in 471 is not a canard, those
Jews left few descendants, so that the Jewish infiltration of
Japan began, for all practical purposes, with the Jews who
crawled in, disguised as Europeans, after Japan resumed intercourse with the West in the second half of the Nineteenth
Century and again after the defeat of Japan in 1945. Although
many of these intruders have intermarried with Japanese, the
total number of Jews and half-breeds in Japan is, he believes,
too small to permit effective subversion and ravage of that
nation by the methods that the Jews have so successfully used

other races produce odorous secretions as by a constant perspiration.
According to Baker (p. 173), almost 10% of the Japanese produce some
odor in armpits; this is regarded as an humiliating disease, which
disqualifies men for military service. It seems odd that a trait so offensive
to Japanese sensibilities has not been bred out of the people over the
centuries. It must come from some Caucasian (White) admixture and is
generally traced to early miscegenation with the Ainu, but it could come,
at least in part, from other sources, perhaps through China, where even
Aryan blood has been absorbed in historical times: one thinks of the
Roman soldiers who made their way to China after Carrhae and, in later
times, the many Europid peoples, some of them unmistakably Nordic,
whose presence in Chinese territory is well attested by the evidence reviewed
by Otto Maenchen-Helfen in *The World of the Huns* (University of
California Press, 1973), pp. 367-375. Of course, an influx of Jews could
account for some of the genetic contamination, but my point is that it
could not be the source of all the physical variations found among the
Japanese and need not be the source of any.

5. This is No. 9 in a series of small booklets, collectively entitled
"Library of Political Secrets," published in English by the Mexican Unión
de Católicos Nacionalistas, some of which have been reprinted in the
United States. The eleven booklets now in print (including No. 9) may be
obtained from Jane's Book Service, P. O. Box 2805, Reno, Nevada, at
$2.00 each, postpaid.

against Europe and the United States.[6] For this reason, Bakony says, the Jews are relying on the "Lost Tribes" hoax to delude the Japanese and undermine the society of "a land Judaism is determined at all costs to conquer and control."

The Jews are therefore promoting a fraud that we may call "Japanese Israel" by analogy with the grotesque fiction which intoxicated many Anglo-Saxons. Bakony even estimates that McLeod may have been more than the simple-minded fantast that he appears to have been:

> McLeod and a number of Japanese professors who, according to my information, are Japanese only on the outside and clandestine Jews on the inside, have disseminated these fables [that the Japanese are descendants of the Israelites and therefore have an "Identity" as Jews] for the purpose of diffusing throughout the country the religious imperialism with which the Jews seek to gain control over the Japanese people.

The attempt to bring Japan under the Jewish yoke by the "Identity" deceit combined with proselytism has had considerable success.

6. Bakony admits, however, that the cuckoos in the Japanese nest constitute a threat to that nation's future. He refers to Japanese authorities who attest that "with the intermarriage of Jewish immigrants (both male and female) from the 19th century on, with Japanese partners, the number of people in the country of Japanese-Jewish descent has been steadily on the rise. They use ordinary Japanese names; they have adopted Japanese customs and even the prevailing religions of Japan, such as Shinto and Buddhism; and they have come to possess racial and physiognomic traits such that it is very difficult to tell them from other Japanese—all of which makes this an infiltration that is becoming extremely dangerous for the future of Japan." In 'Populism' and 'Elitism' (Liberty Bell Publications, 1983), I mentioned obiter (p. 62, n. 40) the terrible success of the Jews in polluting the blood-lines of the British upper classes in preparation for the destruction of Great Britain. In Japan, the work of genetic subversion is even easier, for the Jews are not a White race and, although they may enter the country in the guise of American business men or members of the American Army of Occupation, they have no hesitation in assuring the Japanese of their hatred of the White race that has so grievously afflicted Japan: they are "fellow Orientals" with a racial enmity to the barbarous Aryans, who always "persecute" them. Thus in Japan the cunning invaders can appeal to national patriotism, whereas in England they had to appeal to the Anglo-Saxons' greed and Christian superstitions.

13

"Japanese Israel" is simply "British Israel" with the names changed. Old McLeod's brainstorm was the source of the nonsense in Japan. His prime datum was history that he manufactured by asserting that the first emperor of Japan was named Osee and established his rule in 730 B.C. and must therefore have been the last king of Israel, Hosea (Ōsēĕ in the Septuagint), who lit out for Japan before the Assyrian conquest in 722 B.C. It is charitable to suppose that McLeod came to Japan with a copious supply of Scotland's great beverage. I do not see how he can have failed to know that even if one takes seriously the legends assembled in the Japanese *Kojiki* (of which there is a learned English translation by Professor B.H. Chamberlain) and the chronology that has been attached to it, the first ruler of Japan was Jimmu, great-grandson of the sun goddess, Amaterasu (who, by the way, had the customary virgin birth, but without a Holy Ghost to help with the gestation). Jimmu is explicitly said to be the first human being to govern anywhere in the Japanese islands, and the tradition fixes his date at 660 B.C. (which is why our 1983 is 2643 in the Japanese calendar). Jimmu is, of course, merely a legendary figure and Japanese scholars admit that there is no secure basis for history, as distinct from legends, until ten centuries later. Even the famous Jingo (A.D. 200 or 320) is falsely credited with aggressive warfare and divine inspiration, although she probably did exist and did replace her foolish husband as ruler of some part of a yet ununified Japan.

Jimmu, who was quite literally the Son of Heaven, was traditionally the ancestor of all subsequent Sons of Heaven, including the present Emperor, but if one believes the legends (which are as full of incredible miracles as the Bible) and then shoves a fictitious Hosea into the genealogy, it is as easy to show that the present Emperor is a descendant of David and hence a Jew as it was to show that poor Queen Victoria derived her lineage from the same bandit.[7]

7. It is generally believed that the tales about David were based on the exploits of a Jewish bandit who flourished at some uncertain date and so impressed his contemporaries that he became the hero of a cycle of folk-tales, much as some English outlaw's adventures were elaborated into the stories about Robin Hood, with, of course, the differences that show the contrast between Jewish and Anglo-Saxon mentalities. I note, however, that Dr. Erich R. Bromme, in his *Untergang des Christentums* (Berlin, 1979-80), comes to the conclusion that there was only one David, a captain of the Persian army that kept order in southern Palestine, who

So far as I know, McLeod's balderdash was first taken seriously when it was revived in 1925 by Professor Chikao Fujisawa, who was quickly joined by the Professor Anasaki of whose philological sleight-of-hand we have already seen a specimen. One or the other of them, I believe, produced a real chestnut: the title of the Japanese Emperor, Mikado, is the Hebrew *mi-Gad* and therefore means that he is a descendant of the "lost tribe" of Gad. I am disappointed that no con man in Belfast has had the enterprise to go to Japan and make his fortune by teaching the Japanese that Jimmu is obviously just a spelling of Jimmy, so that the Mikado is indubitably a scion of a Scotch-Irishman who named his son Mike, and since Mike gave his name to his own son, the latter was known as Mike-do, 'do' being the standard abbreviation of 'ditto,' whence the title. And as for great-grandma Amaterasu, who could doubt that her name is simply the Greek definite article (as pronounced in Doric) + the word for 'mother' (= Latin *mater*) + the genitive of the second-person pronoun (misused as in the low Greek of the "New Testament"), which was pronounced *sû?* Obviously Amaterasu means 'your mama.' Jimmy, you see, was a learned Scot, and the Mikados should be proud of such ancestry. Ain't philology wonderful?

I gather that the hariolations of Professors Fujisawa and Anasaki inspired the foundation of a Holiness Church, of which the Bishop, Juju Nakada, proclaims that "it is God's will that these two nations [the Ten Tribes who hit the road for Japan in 722 B.C. and the Two Tribes who have been vampires on the *goyim* in the rest of the world] be united after 3,000 years." And, of course, Japanese Israelites who want to get in on the Holiness will have to turn to and help old Yahweh get his wish. He can't do anything for himself these days, except perform a few trivial miracles in out-of-the-way places when no one is looking.

The Japanese are the politest people in the world, but even so, their failure to guffaw loudly when they hear such stuff

took advantage of the defeat of the Persian Empire by Alexander the Great to set himself up as King of an extemporized Kingdom of Israel in 332 B.C. and disseminated the tales about an earlier David to make his grab of local power seem legitimate. For a summary of Dr. Bromme's conclusions, see Vol. V, pp. 304-307, where the relation of David's forgeries to the Essene-Christian propaganda is stated concisely; for the many passages in which the evidence is presented, see the index at the end of Vol. V, s.v. 'David.'

would be unbelievable, if we did not know that during the past century many literate Englishmen, including a member of Parliament and an astronomer of some distinction, were able to believe the British version of that hokum, and to believe that the reunited Twelve Tribes would make the British Empire eternal. As it is, Bakony concludes his little essay with photographic reproductions of some items from the press. In one of these, from the *Jewish Voice* (17 September 1954), a Rabbi, just back from the Orient, reports that in Japan, a nation demoralized by her defeat at the hands of the Jews' stooges nine years before, "tens of thousands of Japanese men and women...look forward to joining the ranks of Israel."

One does not believe any unsupported statement that emanates from the race that is trying to put over the "Holocaust" hoax, but an item in the *Jerusalem Post* (2 February 1980) seems factual in its report that a manufacturer of paper for computers, who owns a large plant in Japan, is moving his headquarters to the West Bank of the Jordan that the Jews recently took from the Moslems who had inhabited it for centuries. The industrialist, who is the head of a sect of two thousand persons in Japan, claims to be the son of a Japanese general who was killed in action during the recent war. He says that when he was a youngster on Okinawa, he fell seriously ill with tuberculosis and pleurisy, and a Christian missionary brought him a Bible with the usual sales talk. The sick boy read the book and it convinced him that whatever might be said for the Son of God, Papa was still the boss, and that "God had promised *everything* to the Jews and they were his Chosen." And the boy soon convinced himself that he was a descendant of the "Lost Tribes" who had peopled Japan, so that he was himself one of the heirs to everything and that he had better head for Palestine, where the Messiah (Christ) may drop down from the clouds at any moment, in keeping with the Bible Prophecy, to put his heirs in undisputed possession of everything.

The slap-happy industrialist, who bears the odd name of Sadao O'Hara, says that he, as the son of a warrior, is a samurai, and the name of Japan's military caste is a derivative of Samaria, whence they hailed. I do not know what he will do if he ever finds out that the Jews have been working for centuries to exterminate the Samaritans (e.g., their invasions of Samaria in the reign of Claudius around the date that some of the early Christian sects assigned to the Crucifixion) and are now on the verge of success. (There were about 300 Samaritans left alive ten years ago, and Begin has probably found time to cut their throats since then.)

It would seem, therefore, that the Jews are having some limited success in peddling the "Japanese Israel" hoax to weak-minded Japanese. Why a self-respecting Mongolian or Anglo-Saxon would wish to trace his ancestry to the tribe of squalid and vicious barbarian bandits described in the "Old Testament" is a psychological puzzle that defies explanation, but we must accept the fact that some members of both races do have so low an opinion of themselves.

Any nation can tolerate a few thousand eccentrics and oddities so long as the bulk of the population is sound, and if we are to estimate the chances that the Jews will be able to undermine Japan with the "Japanese Israel" hoax, we should take a necessarily hurried look at one episode in the long history of Japan.

JAPAN AND THE WEST

The appropriation of our technology should have occasioned no surprise. It was in keeping with the national character of the Japanese, who, from their first contacts with Europeans, exhibited an extraordinary eagerness to assimilate and emulate our civilization, our techniques, our methods, and even our fashions and fads. It is not too much to say that the Japanese, far more than any other alien race, have been fascinated by our culture, to which they have shown a hospitality that was interrupted only when their equally remarkable sense of racial cohesion and self-preservation made them realize that the advantages of contact with the West could then be bought only at the cost of national suicide.

In the past century, Japan has welcomed our scholars and men of letters, some of whom have reciprocated by so admiring their hosts' culture that they elected to live in Japan. One thinks of the English scholar, Basil Hall Chamberlain, who became Professor of Japanese Philology in the Imperial University at Tokyo. Another prime example will occur to all who have interested themselves in American literature: Lafcadio Hearn, sent to Japan by Harper's to write articles for their magazines, decided to remain in Japan, married a Japanese lady, and became Professor of English Literature in the Imperial University. He eventually decided, for the sake of his children, to become naturalized as a Japanese citizen, and soon thereafter the Japanese government, with a kind of Oriental logic, drastically reduced his salary in the university, on the grounds that a Japanese was worth much less than a European.

Even today, what is fashionable in the West thereby becomes fashionable in Japan, but it is noteworthy that even the most slavish imitations of the West are surcharged with something that is distinctively Japanese. They have taken over our "jazz," but if you listen to a current "hit" produced in a Japanese night club, with the American music played by a Japanese orchestra and the American lyrics sung (usually in both English and Japanese) in the clear voice of a Japanese woman, your ears will tell you at once that the performance is unmistakably Japanese. Some of their artists have unthinkingly imitated the schizophrenic daubs of what some Americans call "modern art" and prize because it reflects the Jewish hatred of all visual beauty, but if you will examine the specimens reproduced on the last pages of Lucille R. Webber's *Japanese Woodblock Prints* (Brigham Young University Press, 1979), you will see that something of the Far East has been added, in color or design, to even the most ugly and repulsive botches. Whatever the Japanese take over from us, they make their own.

The Japanese, who have always been characteristically eager to learn from foreigners, first came into contact with the West around 1543, when some Portuguese, on a voyage from Siam to Macao, were blown from their course and landed on one of the Japanese islands. They were received with wonted hospitality and, significantly, the local ruler, impressed by the crude firearms of the strangers, immediately ordered his armory to find a way to manufacture duplicates of the novel weapons. Thus began more than half a century of mutually profitable trade relations, during which the Portuguese, soon followed by the Spanish and the Dutch, tried to supply the Japanese demand for European wares and brought back rich cargoes of silks and other Japanese goods. Despite the rivalries of the three European nations engaged in this trade, this friendly and lucrative intercourse would have continued uninterrupted thereafter, had it been limited to commercial and intellectual relations.

The Europeans brought with them to Japan syphilis and Christianity. The former was easily kept under control, but the latter soon became epidemic. To the xenophilous Japanese, accustomed to the gentle doctrines of Buddhism and to religious eclecticism by which an individual made his choice among the forty-three popular varieties of Buddhism more or less admixed with Shintō or worked out some compromise of his own, the exotic cult was a great novelty. It differed from the religions they knew as much as (but not more than) "jazz" and "boogie-woogie" differ from the waltz and tango. It was foreign, but everyone knew that the doctrine of the Buddha had come from China, and learned men knew that it had reached China from a yet more distant land. There was a mysterious morbidity in the behavior of a god who had himself killed on a cross to save mortals from the consequences of his own anger, but gods are strange folk. After all, in Shintō, the Creator (Izanagi) had to go down into the underworld to rescue his wife (Izanami), who had died in child-birth, and that was as odd as anything Jesus had done.

Christianity, furthermore, was endued with the great prestige of a race that had attained manifest superiority in building ships, making weapons, and inventing many new mechanical devices and chemical processes, and had an equally great superiority in the knowledge that enabled them to range freely over a vast world in which Japan, Korea, and even China were but relatively small regions: perhaps they also knew more about the invisible world. And Christianity was promoted by Jesuits, who had perfected by long experience the subtle art of adapting their propaganda to the credulity of Oriental races. The new religion thus had a great fascination for the Japanese, and perhaps an even stronger appeal to the *daimyos*, the local rulers who, in that feudal society, were virtually independent monarchs, each in his own territory, and each of whom hoped to secure for his own harbor the profits of a lucrative commerce that was somehow tied up with the exotic religion.

Christianity spread rapidly in Japan, at first without opposition, since the people were accustomed to tolerate every variety of belief about the unknown. But here we are confronted by the question we could not answer above. What about the descendants of the 18,670 Jews and, perhaps, Judaized natives who are said to have been in Japan in 471? Did they leave many descendants eleven centuries later who were Jews, at least in the sense in which Mr. Shi in Kaifeng knew he was a Jew? If so, would they not have flocked enthusiastically to a religion that exalted a Jewish God and a Jewish Saviour? We know, furthermore, that the Jesuits, like the rest of the Catholic Church, were deeply infiltrated by Jews, and it would be a fair guess that some of those Marranos turned up in Japan. Could they have reached some understanding with their surviving congeners in the islands to create the dissension and turmoil from which their race habitually profits? All this, I remind you, is sheer speculation *in vacuo*, with no historical fact to support it.

The history of Christianity in Japan is far too complex to be summarized here, but there can be no doubt about the basic facts.[8] As soon as the apostles of a cult that claimed exclusive possession of the Truth about the Universe obtained a sufficiently large number of converts, they naturally indulged the Christian lust to persecute, destroy, and kill. They incited mobs to burn Buddhist temples, to destroy "pagan" art, to kill Buddhist priests, and to pillage the homes of wicked unbelievers. Feudal lords were induced, by faith or greed, to decree that all residents in their domains must be sloshed with holy water or decapitated. Lords who were so obdurate that they merely extended to their Christians the toleration they were accustomed to extend to all sects found that they had in their territory a tightly organized body of secret subversives, who were zealously exciting mutiny and revolt, and who, as soon as they got a civil war under way, would appear among the insurrectionists with banners that would enable Jesus to see whom he should help and whom he should smite. And, when necessary, his divinity was attested by the booming of cannon aboard Portuguese ships in the harbor.

The Japanese soon discovered that the coveted merchandise and learning of the West brought with it a spiritual plague that menaced their national identity at a time in which they already

8. The most lucid and concise summary that I have seen is in the Eleventh and Twelfth Editions of the *Encyclopaedia Britannica*, Vol. XV, pp. 224-237. There is, of course, a vast amount of writing on this subject in English, much of it more or less vitiated by the tendency of Western writers, even if they are not Christians, to assume that Christianity, though false, is inherently superior to other religions, equally false. For a good presentation of the Christian view, see *The Martyrs of Nagasaki*, by Frederick Vincent Williams (Fresno, California, 1957), which records the claim that after the official suppression of Christianity in 1640, "tens of thousands" of Japanese continued to practice the forbidden cult secretly in the privacy of their own homes, and that there was even a kind of clandestine church that venerated a martyr named Bastian and had some thirty thousand members. These secret cults surfaced after 1865, when Christianity was again tolerated. Whether any or many of the crypto-Christians were also crypto-Jews is, of course, anyone's guess.

had enough troubles of their own. And the chaos was only augmented when Spanish ships brought in squads of Franciscans and Dominicans, who assured the bewildered Japanese that the Jesuits were a pack of vile and perfidious intriguers, while the Jesuits loudly protested the admission to Japanese territory of such scum as the ignorant and plebeian friars. When the Dutch appeared on the scene, it became necessary for the Catholic competitors to agree in assuring the Japanese that Satan had sent the Protestants to seduce True Believers to eternal damnation, while the Dutch lamented the future torments of souls that the fiendish Antichrists in Rome had already snared to be broiled forever on the gridirons of the underworld.

It is needless to remark that the several brands of Christians naturally employed holy mendacity and forgery to win souls and destroy competition. The first Englishman known to have taken up residence in Japan, Will Adams, owed his life to a feudal lord who was too intelligent or bewildered to believe the Jesuits, who assured him that Adams and his companions were pirates who preyed on the commerce of all nations and should be immediately executed. Adams, thus saved, appears to have been an experienced shipwright; he taught the Japanese how to build ships of Western size, and became a trusted friend of the *shōgun* and, in effect, a Japanese nobleman in his own right. It may have been at Adams' suggestion that the *shōgun* despatched a trusted subordinate to Europe to observe the Christians in their own lands, where the amazed and appalled Japanese saw the mighty Westerners engaged in furiously butchering one another to conciliate their ferocious god. The report of that man and of other emissaries finally enabled the Japanese to understand the charms of Christian godliness and the beauties of a blood-thirsty fanaticism.

I will hazard a guess that the Japanese who marvelled at European culture also learned how Jesus had blessed the natives of Mexico and Peru. The Spanish conquistadores were ruthless men, as they had to be, but the Japanese in Europe probably heard tales that had been artfully exaggerated by the Catholic missionaries to those lands, who were always at odds with the civilian governors who refused to obey them. That seemed to confirm the tales told in Japan by each of the three brands of Christians, who solemnly assured the Japanese governors that the other two brands represented Europeans who were planning a sudden invasion and conquest of Japan. The holy men were lying, naturally, but in a sense they inadvertently told the truth, for Japan's narrow islands, surrounded by the sea, were vulnerable, and had European colonies been established there, the innate vigor of our race, not yet palsied by its degenerative diseases, would probably have found a field for action there even before it did in India.

It was in the years around 1600 that Japanese envoys visited Europe, where they had to pretend they were Christians to avoid molestation, and where the several nations of Europe, inspired with Jewish ferocity and righteousness, were helping Jesus save souls and stamp out heresy with bloody ingenuity and zeal as they prepared for the climacteric piety of the Thirty Years' War. The men from Nippon were astounded: they came from islands in which men fought bravely and often cruelly for intelligible and tangible purposes, but the forty-three Buddhist sects disputed vigorously about what the Buddha had meant while showing a polite consideration for each other and even good humor. The envoys, being benighted pagans, failed to see the need to help Jesus annihilate the Antichrist and all his minions, but they probably saw something of the deadly efficiency of European armies and returned home with a prescience that Japan had only the choice between becoming (anticipatorily) another Tahiti or finding some means of preserving her national identity and culture.

It would be tedious even to adumbrate the tangled and confusing events of the next forty years, as the Japanese in political power fluctuated between their desire for foreign trade and the arts and learning of the West and their fear of the demoralizing religion that Europeans brought with them as a deadly infection. In the meantime, as had happened earlier in Europe, the feudal system declined, and the successive Shōguns, who ruled in the name of the unapproachable Mikados, extended their authority of the central government over the whole of Japan. Eventually, Christianity was exempted from the toleration accorded all other religions and effectively suppressed, in blood where necessary. The Catholics were expelled, at first to the great profit of the Dutch, who for some years enjoyed a most lucrative trade on the condition that they would not attempt spiritual subversion, but finally even they were almost squeezed out.[9] After 1641, Japan embarked on a

9. The Japanese felt in honor bound to observe the letter of their treaty with the Dutch, which they never repudiated, but they imposed restrictions that the treaty had not explicitly forbidden, eventually confining the Dutch traders to Deshima, a tiny island in the harbor of

policy of total isolation, refusing admission to foreigners and forbidding Japanese to go abroad to lands whence they might bring back the plague.

This policy has naturally aroused supercilious or indignant comment from Aryans who believe that Japan should not have protected herself from a degenerative disease by what seems a kind of obscurantism. I need only quote an authoritative summary.

> It is apparent that . . . Christian propagandism was responsible. The policy of seclusion adopted by Japan in the early part of the 17th century and resolutely pursued until the middle of the 19th, was anti-Christian, not anti-foreign. The fact cannot be too clearly recognized.[10]

The policy was maintained until the Japanese were convinced by events that our technology had so advanced during the intervening two hundred years that it was no longer possible for Japan to resist invasion.[11] And from that fact the Japanese intelligently drew the lesson that they must learn and appropriate the technology against which they had become defenceless. How brilliantly they did so is now apparent, when, after having been terribly defeated in war, they are now defeating us on *our*

Nagasaki, where they were kept in a kind of quarantine, and no Japanese (except prostitutes) could visit them without an official permit from the local authorities. Under these onerous restrictions, an exiguous trade between Holland and Japan was continued through all the years of Japan's otherwise total isolation.

10. I quote from the article in the *Encyclopaedia Britannica* cited above.

11. A few years ago I heard a university lecturer unctuously tell an unprotesting audience how "peacefully" Japan had been "opened up to Christian civilization" by Commodore Perry and his fleet of steam-powered warships in 1854. It is true that the Japanese, overawed by the cannon of Perry's ten "peaceful" warships, made some concessions, but Japan was not really "opened" to foreign commerce until after a British fleet had bombarded the city of Kagoshima and reduced it to rubble, and another British fleet, with a few American, French, and Dutch vessels added to make it seem international, levelled Shimonoseki in 1864 and imposed a fine of $3,000,000 on the local population. That finally convinced even the most reluctant Japanese of the charms of Western civilization, and thereafter they set out whole-heartedly to acquire its technological power.

own terms.

There is one question that we must again ask and be unable to answer. The suppression of Christianity in Japan necessarily involved bloodshed on a large scale. Western writers are usually most distressed by the fate of Christian missionaries who, having been expelled from Japan and courteously warned not to return, sneaked back into the islands under various disguises to continue their work of subversion until they were finally apprehended and executed. They must be presumed to have won the reward they sought. What concerns us here is the actual depletion of the Japanese population by the execution of Japanese who had become obstinately infatuated with the exotic religion. It seems impossible to estimate the total number slain. We hear that the Jesuits alone had made some 300,000 converts by 1595, and of many thousands brought to Jesus at various times and in various regions thereafter. How many of these recanted during sporadic efforts to restrict or suppress Christianity in one region or another is uncertain, as is the number who, in the end, like Panurge, maintained their convictions *jusqu'au feu exclusivement.* We should not reckon as Christians all the persons who perished on the losing side in revolts and civil wars that were incited by the Christians or in which Christians largely participated. The commonly quoted figure of 235,000 "martyrs" is considered by some writers excessive, while others would accept 500,000 as possible. We cannot even guess how many Jews who looked like Japanese perished in the several massacres of Christians, nor yet how many Marranos there were among the Japanese who are said to have clandestinely continued their practice of the forbidden cult after it had been officially suppressed.

The Japanese have shown a remarkable ability, for which I can call to mind no historical parallel, to throw off a foreign infection. One could even speculate that their experience with Christianity, like recovery from certain epidemic diseases, may have been actually beneficial, imparting a certain immunity by strengthening their racial consciousness and sense of national unity. It seems likely that "Japanese Israel" will have no significant effect.

JAPAN'S FUTURE

If Japan is deeply infested with Jews, she is doomed, and the details of her ruin would be of no interest, even if we could foresee them.

24

If she is not, she has a formidable potential, and unless she is destroyed by some external force, she will determine some part of the future of life on this globe. Her people have shown a national genius that gives them incontestable superiority among Mongolians. In the 1930s she tried to dominate and organize her race, and if she has an opportunity to do so, she will try again. Speculations about the unpredictable century ahead of us must include the obvious possibility that the rotting of our race will continue and the nations of the West will perish in the ignominious paresis they have brought upon themselves. The future will then belong to the Mongolians, and if they are directed by the manifestly superior intelligence of the Japanese, they will own this planet. Some of our descendants will probably survive to become the Ainu of Europe and North America.

The only other people who can match the Japanese in racial cohesion and a high average of intelligence are the Jews, but they are a parasitic race and have shown throughout history only a terrible and highly specialized power to destroy, to suck the life out of the nations on which they have fastened themselves. They have never shown the slightest ability to establish a civilization or even a viable barbarism of their own. If our race perishes and they cannot transfer themselves to the Mongolians, they will perish with us. Some of us may find that thought consoling.

Conjecture about the future, even if it is realistic and rational instead of romantic and emotional, can suggest a wide variety of possible consequences of the present, and too often the course of history has been turned by events that were unexpected and beyond human calculation. It is possible that our race has a latent vitality that will enable it to recover from a disease that now seems mortal. It is also possible that the Japanese have some latent infection or organic weakness that has not yet become apparent.

Very few men of our race have mastered the intricacies of the Japanese language and of the modalities of thought that are far greater obstacles than grammar and vocabulary to one who would read and understand the voluminous annals and litera-ture; and of the few who have done so, yet fewer can appraise dispassionately and objectively what they have read. They alone are entitled to speak of the soul of Japan or, rather, of so much of that soul as can be perceived by an Aryan mind. They can at least measure the impassable gulf that separates the two races;

the rest of us can only estimate it superficially.

One could compile an enormous bibliography of books about Japan in Western languages, even if one restricted it to books worth reading. There are many in English, more in German, a large number in French, and some in Spanish and Italian. I shall mention only the political histories by Murdoch & Yamagata and by Brinkley & Kikuchi, the history of literature by W. G. Aston, and the many interpretative essays by B. H. Chamberlain and Frank Brinkley. There is literature on Japanese subjects, notably the finely wrought prose of Lafcadio Hearn and John Masefield's memorable tragedy, *The Faithful.*[12] There are many translations from the Japanese. The earliest books in the language are compilations of traditions and legends from a prehistoric past, and these have been ably translated, the *Kojiki* by Chamberlain and the *Nihongi* by Aston. Donald Keene has edited an *Anthology of Japanese Literature . . . to the Mid-Nineteenth Century*, and there is even a *Book of Japanese Verse* in the Penguin series.

To my mind, the soul of Japan may be descried through a shimmering veil of impressionism in the Nō plays, of which there are more than five hundred, about half of which are still performed at times. There are several selections in English, but I think the misty melancholy of the more poetic plays was best rendered in French by Steinilber-Oberlin & Kuni Matsuo (Paris, 1929). Of what we may call the legitimate drama, the most famous author was Chikamatsu, whose plays have been translated by Donald Keene. I know Tarahiko Kori's heroic tragedy, *Yoshitomo*, only in the Spanish version by Antonio Ferratges (Madrid, 1930).

The best-known Japanese novel is, of course, Lady Murasaki's *Genji*, translated by Arthur Waley. Modern novels, written in imitation of Western fiction, must be used with great caution: some are trash. Yukio Mishima's *Forbidden Colors* depicts painfully the demoralization that followed the defeat of Japan in 1945, from which she so quickly and brilliantly recovered. A

12. I need not remark that while *Madame Butterfly* is a beautiful and moving opera, it tells you very little about Japan. It was based on an English imitation of Pierre Loti's *Madame Chrysanthème,* which has some value, especially for the author's realization that he and his temporary wife had mentalities so diverse that, despite their domestic intimacy, no reciprocal understanding was possible. Cf. the experience of an American officer that I shall report below. It turns, of course, on the fact that he was American, not French.

better treatment of the same subject is Keene's translation of Osamu Dazai's *The Setting Sun*, an almost cruel portrayal of the demoralization, but accompanied by an indication of the reason for Japan's resurgence: the man who does not give way to despair perceives that the Western poison, egalitarianism, is "the obscene and loathsome vengeance of the slave mentality." A nation which does not forget that fact of life in its darkest hour is a nation that has the potentiality of greatness.

A people is best known through its myths and literature, but we obtain some further information, and a vast amount of misinformation, from Americans who have served in the army of occupation with which we afflicted Japan or were stationed there at the time when the government in Washington seized an opportunity to get many young Americans killed in Korea, to abort a possible prosperity in the United States by inflicting more taxes on the serfs, and to convince the whole world that Americans are utterly contemptible.[13]

The most acute of our observers return with one fundamental datum: "The Japanese are so polite that you can never even guess what they are thinking." They are so polite they speak English to relieve the Americans of the need to acquire even a smattering of colloquial Japanese, although some may learn

13. When we look back on that ignominious episode, we are apt to forget the subtlety of the propaganda that was employed in that application of the principle, "perpetual war for perpetual peace." Intelligent observers, naturally, were unimpressed by the mouthpieces in the White House and elsewhere that yammered about "resisting aggression," supporting the "United Nations," and similar bilge, but another version had been prepared for them, a "confidential" report that the United States was belatedly moving against the Soviet Empire, and that the "real purpose" of the meddling in Korea was to "escalate" that action into a general attack on Bolshevism throughout the world and especially in its heartland, Russia. That seemed plausible at the time and for months or even a year fooled rational men. The ranking officer in Military Intelligence, whom I shall mention in the second paragraph below, was taken in by that clever deception when he was sent to Japan to direct from headquarters there certain kinds of espionage in Korea. He said that it was four months after the beginning of hostilities before he began to suspect the horrible truth, and as long before he saw that there was *no* alternative explanation—despite his access to much information that was concealed from the public. So great was then the weight of the old military tradition that American armies should try to be victorious! Some vestiges of that tradition even survived as late as the time when Vietnam was selected as a fresh pretext for bleeding the boobs.

enough to add a few words to current slang.[14] Some men return with a Japanese religion[15] or a Japanese wife.[16] But few learn more about Japan than do tourists who spend two or three days ashore from a cruise ship.

What Westerners see and learn in Japan is strictly limited by the innate character of the Japanese, who instinctively combine courtesy with an inner reserve that is foreign to our nature. They keep their thoughts private, secret, with a mental discipline of which our race is incapable. I shall give you a good illustration of that basic fact.

When the United States was preparing to start shooting in Korea, a ranking officer of our Military Intelligence was sent to Japan to supervise from his headquarters there certain intelligence operations in Korea. He provided himself with a Japanese mistress from a very good middle-class family. His cover was some position in the Quartermaster's Corps, and, of course, he was careful not to let his concubine know that he had any other military function. She was, in every way, a perfect mate, who seemed to anticipate his every wish and desire by some kind of instinct, sometimes when he was scarcely aware himself of

14. One of the vulgar euphemisms in current use is 'fox,' which designates a young and especially libidinous female, being applied in college circles to the 22% of "co-eds" who, according to a recent survey, are eager to copulate on sight with any presentable male. In Western tradition, the fox is a type of craftiness, not lust, so I conjecture that this use of 'fox' (not, notice, 'vixen') had its source in Japan, where a particularly liberated and lascivious "sporting woman" (*asobime*) is said to be a werefox or to have been created from the bones of a horse and animated by the fox-goblin; hence many allusions in satirical verse.

15. The religion is most commonly the Rinzai sect of Zen, one of the sixty or more varieties of Buddhism that the Japanese have fashioned in their own image out of the Chinese Ch'an, which was a radical Chinese revision of the religion that was fashioned in India out of a travesty of the philosophy of Gautama, the Buddha. The Japanese doctrine is described by Alan W. Watts in *The Way of Zen* (New York, 1957) as well as any unverifiable belief can be described by a believer. Zen is said to be the basis of the warrior's code of *bushidō*, but I suspect that the connection is as adventitious as the relation of the Western chivalric tradition to Christianity.

16. An American technician tells me that he is happy with his second or third wife, whom he brought back from Japan. "In Japan," he says, "females are still women." I leave to the reader reflection on the social implications.

precisely what he wanted.

The young woman's two or three brothers had been officers in the Japanese army and had been killed in action. Her uncle and aunt had perished during our fire-bombing of Tokyo, when we destroyed sixteen square miles of the city and made a million persons homeless after a hundred thousand had been burned alive, boiled in the canals, or suffocated by the fire-storm. She and her mother had barely escaped alive from their burning home. The American tried to discover what his perfect concubine was really thinking, so, after many months of conjugal intimacy, he asked her about that American raid on Tokyo. Oh, yes, she remembered it vividly: she had seen the American planes come in like celestial butterflies, "silver wings in the moonlight, very pretty, very pretty!" It was only then that the American, being a highly intelligent man, realized how she hated him—hated him with an implacable—and noble—hatred.

Whether she was watching him for a Japanese intelligence service (their maid, who did not live in, but came every day, would have made a perfect messenger), the American never knew, but that did not matter. Being, as I have said, a highly intelligent man, he realized that he had glimpsed for a moment the soul of a great race.

A nation may also be known by its deeds. We are Aryans, who can think dispassionately; we can recognize great achievements and we can salute with respect brave and gallant enemies. Hitler's recognition of the Japanese as "honorary Aryans" was not merely the slick verbiage that is used in diplomacy to grease a temporary alliance for a common purpose. It had a basis in hard facts. The Japanese had shown, alone among all other races, a quality of mind that enabled them not only to assimilate the science and technology that was uniquely the creation of our race, but also to carry on and augment our work. Writers who are alarmed by Japanese achievement today would be astonished to see how much of what they say was anticipated in 1936 by Anton Zischka in his *Japan in der Welt, die japanische Expansion seit 1854*, published by Goldmann in Leipzig and widely circulated in Germany, although it openly challenged much of Hitler's policy.[17] It would have been

17. Zischka used the phenomenal achievement of Japan as a basis for a study of the relative efficiency of the *laissez-faire* economy of European nations as contrasted with the nationally unified and directed economy of

irrational to deny that the Japanese were an exception to many generalizations about non-Aryan peoples.

The term "honorary Aryan" recognized a similarity that is moral no less than intellectual. Whatever the racial explanation, and despite the great genetic differences, the Japanese are the alien people whose moral qualities most surely command our esteem. Although *bushidō*, the code of the warrior caste, differs in several respects from our chivalry, it is a high standard of personal honor, such as our race instinctively admires and prized highly before our manhood was rotted by a spiritual leprosy. We, no less than the Japanese, spontaneously admire the Forty-Seven Ronin, whose devotion is celebrated in John Masefield's *The Faithful;* their heroic loyalty reminds us of what Tacitus tells us about the *comitatus* of a Teutonic chieftain or of the words and deeds of Wiglaf in *Beowulf.* Our native Aryan (as distinct from Christian) morality lifts our hearts when we hear of brave men to whom the knightly virtues mean more than life itself. And although the practice of *seppuku* seems to us gratuitously and extravagantly ascetic, our racial psyche still knows that death wipes out dishonor, and that no right is more indefeasible than a man's right to his own life, which can be limited only by a duty that he has assumed and must honorably discharge before he is free to dispose of what is inalienably his.

We must admire the great accomplishments of Japan in her stupendous, though premature, effort to conquer for herself a great empire in Asia, and the energy and valor with which she fought in the Pacific War we forced on her. It is true that prisoners of war were treated with great cruelty, but the Americans, who have repudiated all the conventions by which civilized nations of the West tried to attenuate the horrors of war, are in no position to complain about that; they should

---

Japan. He made almost no allowance for innate racial differences, but he did wonder why our nations had permitted the rise of Japan, and he more than hinted that our whole race even then faced a crucial decision that might determine its whole future. Of the problem of economic organization he said, "Daß wir es *rechtzeitig* studieren, mag über Lebensfragen der weißen Rasse entscheiden." (His emphasis.) He did not foresee the war against Germany that traitors in Britain and the United States were then planning, but he did see that Soviet Russia and the United States were in fact allied against Japan, trying to encircle her, and might attack her in the near future.

instead take what satisfaction they can from the precedents they have set: in the next war there will be no prisoners.

We can only salute with awe and envy the national and racial devotion exhibited by Japanese soldiers, especially the men who went to die alone for their country and people. The heroism of the *kamakaze* has been reported in a new book, *The Sacred Warriors*, by Mr. and Mrs. Dennis Warner and a Japanese naval officer, Sadao Seno.[18] It deserves to be read with a solemn appreciation.

But let there be no mistake. The Japanese are an admirable people, but they are, and always will be, our enemies, although it is not inconceivable that in some not unimaginable future contingency they and we may be allied against a common enemy until he has been defeated to the satisfaction of one or the other ally, who will then dissolve the alliance as light-heartedly as the American colonists, at the end of their War for Independence, betrayed the French, who had done so much to win that war for them.

There can never be amity between Aryans and Mongolians: the racial chasm is too deep for that. But we have made ourselves the special object of Japanese animus. The series of articles in the *Wall Street Journal*, to which I referred at the beginning, included (11 October) a squawk by a journalist who learned that the Japanese are discreetly eliminating from their textbooks the lies and drivel they inserted at the behest of their conquerors in 1945, and we may be sure that the best minds, at least, are thinking more deeply.[19] They must know that Japan was tricked into her desperate attack on Pearl Harbor so that the unspeakable creature in the White House could please his Jewish masters by stampeding hordes of crazed Americans

18. New York, Van Nostrand, 1982. The book opens with an incident that is probably historical: a Japanese submarine was about to shell San Francisco on the night of 24 December 1941, but was forbidden to do so by the Japanese Admiralty on the grounds that Christmas was a sacred holiday in the United States.

19. Professor James Martin informs me that his sources show that the English translation of Hiroyuki Agawa's *The Reluctant Admiral* [*sc.* Yamamoto] omits important passages in the Japanese original (Tokyo, 1969) dealing with American espionage and advance knowledge of Japanese plans. Whether or not the omitted passages are accurate, they illustrate the kind of thinking a Japanese could discreetly put into print in 1969.

against Germany.[20] They must guess that the United States' secondary purpose in the Pacific War was to strengthen and magnify the Soviet in preparation for the time when, as official Washington realized in 1945, the defeat of Germany made necessary the rapid creation of another world power with which to scare and bamboozle the taxpaying boobs in the United States. And the Japanese will never forget the nauseating obscenity of the farce that the white barbarians staged for more than a year in Japan as a pretext for murdering some of Japan's greatest soldiers, primarily to provide a counterpart for the more savage murder of Germans and a sounding board for the propaganda in the minority report of the kangaroo court, which deprecated lynching of the Japanese generals and admirals on the grounds that they had not been guilty of Germany's awful "war crimes" (i.e., disrespect for God's Holy Race, which, of course, is what is meant by the frantic repetition of the preposterous hoax about a "Holocaust").[21] The Japanese under-

20. The way in which the trick was worked is briefly described in *America's Decline*, p. 7. Japan attacked Pearl Harbor in a desperate effort to avert a surprise attack by the Americans, which she had been led to believe imminent. While the means by which the Japanese were deceived are still little known, it is now established beyond all possible doubt that the loathsome creature the Americans had elected as their President designedly sacrificed the American fleet at Pearl Harbor to ensure a squandering of American lives, resources, and national honor in the war against Germany that he had successfully and secretly incited in Europe, with the help of a British traitor named Churchill. I need not say that the widely circulated book, *At Dawn We Slept*, published under the name of the late Gorden R. Prange, is just a part of the flood of whitewash that is constantly spewed out by venal publishers to keep the Jews' cattle from becoming restive. According to the review by Roger Pineau in the *Christian Science Monitor* (7 December 1982), of which I owe a copy to Professor Martin, Prange's work could not be published until it had been revised by Goldstein and others. It is quite possible that Prange's work, which was ready for publication in 1963 but withheld from the press until 1982, had to be censored to eliminate admissions that might have jarred the brains of Americans sufficiently to start the mechanisms of thought.

21. It is amusing that God's Own are beginning to quarrel over the best ways to keep the American boobs so befuddled that they will continue to believe the great Holohoax. The American Jewish Commission on the Holocaust, headed by the Arthur J. Goldberg who was for many years a Justice of the Revolutionary Tribunal that Americans still call their Supreme Court in Washington, ended in a glorious row in August 1982, although the fact was kept secret until it was disclosed in the *New York Times* on 4 January 1983 (see also the *Christian News* for the week of 17 January). The altercation became so bitter that one Jew accused another

stand and will always remember what they are now too polite to tell us to our faces.

They doubtless regard us with contempt that they will politely and patiently dissemble until their territory is freed of our odious presence. Had we defeated them in a war with a rational purpose, the reëstablishment of our race's dominion over the planet or the annexation of territory in Asia for military bases or colonies, they would respect us. Such doubts about us as they retained after our victory were answered by the acts of national insanity in Korea and Vietnam, which convinced impartial observers that the United States is populated by a horde of madmen who will soon destroy themselves. The Japanese seem to retain some respect for the Europeans, whom we also ruined in one of our outbursts of frenzy. I am reliably informed that a Japanese professor is now discreetly making studies in Europe to determine the causes of the mental disease that has made European nations admit to their countries the Mongoloid "boat people" from Southeast Asia and other

---

of a "desire to write revisionist [i.e., truthful] history." The brawl, according to the press, was caused because some of the Jews wanted to disclose to the public the fact that the Jews in the United States, while yowling in 1939 to 1942 for war to punish the Germans for disrespect to God's Race, seemed unperturbed that the wicked Germans were then exterminating 20,000 Jews a day, as the Holohoaxers would have us believe. Of course, the Jews were then unperturbed because their fellow tribesmen were swarming into the United States from Germany at a very satisfactory rate and naturally knew nothing about the great hoax, which seems to have been invented in December 1942 and was then quite obviously "atrocity" propaganda, such as had been used to excite the American cattle for their Holy War against Germany in 1917. By the end of 1942, the State Department in Washington had officially and in defiance of American law imported half a million Jews and many more had sneaked across the borders from Canada and Mexico or been landed surreptitiously from small boats at various points on the Atlantic seaboard or even from quite large ships on Long Island, a favorite point of entry, since the disembarking invaders could conveniently be met by the limousines of their wealthy kinsmen from New York City, while the American Immigration Service was compelled by a Presidential order to look the other way. The Jews, of course, have shown great restraint in claiming that only 6,000,000 of the Holy Race were exterminated in Germany. According to the *Babylonian Talmud,* the wicked Romans in A.D. 135 slew 800,000,000 Jews in the town of Bethar alone, and the flood of sacred blood thus shed was so great that it rolled forty miles to the Mediterranean, carrying huge boulders with it, and stained the sea red for four miles from the shore. The town of Bethar had an area about equivalent to six blocks of an American city.

potentially hostile aliens, often in the guise of "refugees." To the Japanese mind, such folly is explicable only on the basis of some psychotic malady that has become epidemic. The Japanese scholar's investigations may lead him to the conclusion that is implicit in Jean Raspail's prophetic novel, *The Camp of the Saints.*

We may now return to the question, Will the Jews act in the foreseeable future to diminish or destroy Japan's ever increasing industrial supremacy?

So far as we can now determine, Japan seems proof against the standard techniques of internal subversion. She gave in the past an impressive demonstration, unparalleled in recorded history, of an ability to throw off a deleterious superstition, and there is no reason to believe that she has lost the will to maintain her national health. She is obviously immune to the "one world" fever and the delirium it induces. She has, naturally, a few stupid or degenerate "intellectuals" who ape Occidental nonsense and think that wagging their tongues is evidence of cerebral activity, but they seem to be harmless pests. According to present reports, the Jews' favorite device of inciting class-warfare and implanting the "social justice" disease seems likely to have no appreciable effect. The realistic attitude toward sex, *shunga*, would make futile any attempt to attack the Japanese through pornography, sexual "liberation," and induced depravity. Hallucinatory drugs become a drug on the market in Japan, for the vigilance of the police is made largely unnecessary by the good sense of the majority of the people. And we must remember that, as the London Correspondent pointed out in his article, the *average* of intelligence is much higher in the Japanese population, which is not afflicted by the Western urge to breed out mentally superior strains in its own race in the interests of universal mongrelization and pious idiocy. In fact, it would seem that all the techniques for undermining and demolishing civilization that are in use in Occidental countries are ineffectual when applied to Mongolians who are immune to Christianity.

Japan, needless to say, is now terribly vulnerable to war, and could be crushed between the Soviet Union and the United States, which still has the military capacity to give the Soviets quite effective support. Thus far, however, there are no indications of a design to stir up the Americans for another Holy War. A bundle of very low-grade journalism called *Parade*, which is distributed with the Sunday issues of many newspapers, has carried at least one article on Japanese "inhumanity" to prisoners, but that seems to have been just a foil for the incessant defamation of the Germans, whom the Jews still hate even more than they hate other Aryans. Use of our remaining army and navy for a serious military purpose would, of course, have to be preceded by action to purge the armed forces of the niggers who now clot them and would paralyze them in the event of hostilities. And it would require time to induce the requisite hysteria in a population now festering with peace-lubbers, rambunctious females, epicene perverts, and drug-addicts. Of course, both the Soviet and the United States possess nuclear weapons that could be used to devastate the Japanese islands and require only crews of technicians, and the Soviet, furthermore, also has the biological weapons it is using in Afghanistan and perhaps others equally lethal—it would be hard to imagine anything *more* deadly than the agent that is called "Medusa," because it freezes its victims to instant rigidity in death,[22] and another, which induces an incurable leprosy. These and the three other biological weapons (which permanently incapacitate rather than kill) that are being tried out in Afghanistan could be applied by a comparatively small number of aircraft. So far as is known, Japan would have no defense against such technological warfare.[23]

22. This is probably the weapon that was tried out in a remote part of Siberia almost twenty years ago and observed from a distance by a British spy, as reported by Kenneth de Courcy in his *Intelligence Digest* before he was silenced. A large plane flew over the test area and seemed to scatter particles of an almost incredibly potent cryogenic chemical; the animals and human beings staked out in the area were instantly frozen. The intense cold seemed to be limited to the test area, and the distant observer could not see the particles fall, so that it is possible that the effect was produced in some other way, e.g., by a kind of radiation, conceivably a beam of neutrons accompanied by some other sub-atomic emanation. According to the reports that I have seen, the victims in Afghanistan are found frozen to cataleptic rigidity with such suddenness that their stiff fingers are on the triggers of firearms they did not have time to discharge, but it is not stated whether or not intense cold was observed.

23. Japan has, of course, the capacity to develop nuclear weapons of great power, and as long ago as 1962 members of the government stated that the production of such weapons "for strictly self-defensive purposes" was not precluded by the settlement that the American conquerors forced on the nation. In 1967, when gullible persons in Western nations were excited about "negotiations" for a scrap of paper called a

THE CHINESE PUZZLE

A projection of Japan's future must include some estimate of the likelihood that she can obtain control of the enormous masses of China. I shall not hazard a guess about what she may eventually accomplish, if she survives for four or five decades. Even the present status of China is sufficiently uncertain. We must also consider the possibility that China, which has been equipped with an arsenal of nuclear weapons by persons, thought to be Chinese, who were trained in the United States for that purpose, could be mobilized to join the Soviets in an attack on Japan, should the latter's growing industrial supremacy alarm the powers that decide such matters.

Unlike Japan, China is a land in which there is abundant evidence of deep penetration by Jewish contingents over the past two thousand years, and there is every reason to believe that the Jews are today far more numerous and powerful than the report in the *Daily Telegraph* about the Jews in Kaifeng would suggest. Itsván Bakony in a little booklet, *Chinese Communism and Chinese Jews,*[24] has conveniently assembled, from official Jewish publications in English and Spanish, some of the evidence of continued infiltration of China since the first century B.C. For example, a contingent of Jews arrived to join their fellow tribesmen in the Twelfth Century, and thereafter, at least, the Jews in China prospered mightily. It is a reasonable supposition that, as is the habit of their race, they spread subversion and had some part in the revolt that established the Ming Dynasty (1368-1644), under which Jews climbed to high offices in the Imperial Government, holding positions as commanders of armies, governors of provinces, and Prime Ministers. They were frequently reinforced by fresh bands of God's Pets, chiefly from their large colonies in Persia and India, and we may be sure they played their part in the almost total

---

"non-proliferation treaty," the Foreign Minister of Japan (Miki) denied that his country was equipping itself with nuclear weapons, but reaffirmed the right to do so. The extent to which Japan has equipped herself with such weapons is unknown, but even if her capacities were very great, how could she defend her small and densely populated territory from missiles launched from Soviet or Chinese territory across the narrow Sea of Japan?

24. This is a booklet in the series, "Library of Political Secrets" cited above, as is the booklet by Bielsky that I shall shortly mention. I refer to these booklets, in preference to more elaborate sources, because they are so generally and inexpensively available.

corruption of government and demoralization of the native population that led to the collapse of the Ming Dynasty.

In the early days of the Ch'ing (Manchu) Dynasty, which ruled China from 1644 to 1912, the more prominent Jews scuttled to Shanghai and Hong Kong, where they were protected by the simple-minded British and speedily acquired a virtual monopoly of the trade in cotton and opium. Since the Manchus seem to have felt no veneration for the international race, it is a fair inference that the Jews who had been established in China long enough to acquire Mongolian features and Chinese names found it profitable to masquerade as Chinese and make a secret of their tribal organization. Many of the Manchu Emperors tried to check the corruption of the nation they ruled, and everyone knows that Great Britain declared war on China in 1839 to compel the Emperor to permit the importation and sale of opium for the profit of Jews who had embedded their mandibles in Britain and India, including the famous and wealthy family of the Sassoons. Their seemingly Mongolian kinsmen in China doubtless shared in the loot. Another assault on China in 1856 was necessary, however, to remove the last obstacles to the importation of opium and Christianity. The latter was of no great consequence and seems to have been used primarily to make superstitious English women simper in anticipation of meeting hordes of Mongolians in Heaven, thus neatly preventing Englishmen from questioning the morality of a war to force a whole nation to become addicted to a poisonous narcotic. Thousands of missionaries, both Catholic and Protestant, overran China but were able to talk or bribe only a comparatively few Chinese into "conversion." Some observers have estimated that at any one time only some 18,000 Mongolian souls were ready for dispatch to Jesus.

The two Opium Wars raise a crucial, but unfortunately unanswerable, question. Did the Jews use their British troops to force opium on China merely for profit, or did they derive a spiritual joy from making even Mongolian *goyim* addicts of an expensive drug that both paralyses the mind and will and ruins so many victims economically that all but the wealthy must resort to crimes of violence to meet the expense of stupefying themselves? It is true, of course, that in all societies disintegrated by an increasing rate of crime, as in the United States, the result is highly profitable to parasites, but if we knew the answer to the psychological problem, we could predict with some confidence our immediate future.

Large contingents of Jews continued to arrive in China after 1856, but Bakony notes especially the power of the Soong family of Jewish bankers, who had acquired the outward appearance of Chinese. One of the daughters married Dr. Sun Yat-sen, while another married Chiang Kai-shek. Both of these female firebrands look like Chinese and were graduated from Wellesley College in the United States. Although Bakony does not say so, anyone who inspects a good photograph of Sun Yat-sen will see that the man was not a pure Mongolian, and will suspect that the other side of his heredity went back to Abraham. However that may be, Sun Yat-sen sweated at every pore ideals taken from Karl Marx, and whatever his intentions may have been, his subversion of the Chinese Monarchy initiated half a century of civil war and anarchy during which the Chinese population was the prey of an almost endless succession of enterprising bandits who called themselves "war lords" and repeatedly devastated every region of China not under Western Rule.

One of the most successful of the "war lords" was Chiang Kai-shek, who looked like, and may have been, a real Mongolian and seems throughout his life to have been directed by his wife, Soong Mai-ling. After he broke with the Communists, he proved himself a competent general and eventually ruled most of China until the Americans, with the treachery for which they are now famous, betrayed him to the Soviet-controlled Communists and he barely escaped to set up a government in exile on the island of Formosa, often called Taiwan.

Needless to say, the rulers of the United States, having skillfully arranged to deliver mainland China to the Communists and established a dissident régime on Formosa, thus obtained an almost endless supply of "problems" to distract the minds of their tax-paying serfs and to provide entertainment in the great vaudeville show at Forty-fifth Street and East River, commonly called the United Nations. Good shepherds know how to herd and fleece their sheep in a "democracy."

The number of Yellow Jews, who can masquerade as Mongolians, in China at this time has been variously estimated. Bakony thinks they are not more than 2,000,000, a small fraction of the total population, but we all know that a comparatively small number of Jews will suffice to capture and destroy a nation.

There are only two facts about Communist China that need

concern us. As soon as the new Communist régime was securely established, the widow of Sun Yat-sen, the Yellow Jewess Soong Ching-ling, popped up as the second most powerful individual in the country and was probably prevented only by her sex from becoming the actual head. She must have been the figure about whom her fellow tribesmen rallied, and she remained influential until her death, but she was gradually eclipsed by Mao Tse-tung, one of the cleverest of the bandits who flourished in the 1920s.[25] He had been slowly climbing to the pinnacle of power, which he attained around 1968 and, with some vicissitudes, retained until his death in September 1976. Bakony believes that Mao was a real Chinese, who knew how to use the Jews for his own ultimate purposes.

Despite the racial difference, which precludes any real amity between Mongolians and Slavs, the Communist government of China for years remained on the best of terms with the Communist government of the Soviet, which the United States was constantly shoring up and partly financing, but a rupture occurred not long after the death of Dzhugashvili, alias Stalin. Mao vehemently objected to Khrushchev's denigration of the dead hero, which, although couched in the usual double-talk of politicians, was largely motivated by the fact that Stalin, near the end of his life, had taken some measures against the Jews in the Soviet, most of which were aborted by his sudden and oddly opportune death. Mao's opposition was, naturally, expressed in the usual gibberish of politicians, and no one knows to what extent it was motivated by a wish to deal with the Jewish problem in his own country.

Bakony was convinced that Mao's policy was inimical to the Jews and that the international race would, through the Yellow Jews in China, incite revolts against him. This could be the explanation of several abortive attempts to overthrow him.

---

25. Mao's rise from bandit to ruler is described by George Paloczi-Horvath in *Mao Tse-tung, Emperor of the Blue Ants* (New York, Doubleday, 1963). The author reads and uses Chinese sources, and although basically hostile to Mao, tries to be objective. He ends his book with the odd suggestion that if the Soviet-American combine were to give Mao a hug and a kiss, he might become a good boy. Many of our would-be intellectuals read Mao's voluminous writings in English translations and rack their brains in an effort to understand his version of Communist "theory": they have not discovered that Mao, whatever his race, had the distinctive habits of the Jews and, like them, used words, not to communicate his thoughts, but to conceal them.

Bakony wrote before Mao publicly stated that while scattering the blossoms of social justice and brotherhood, he killed 800,000 Chinese. All observers believe that the great man modestly understated his accomplishment, and some estimates of the total number of his victims are as high as 12,000,000. His social engineering certainly gave him an opportunity to solve China's Jewish problem with the Oriental practicality of which our race seems incapable, but, so far as I know, there is no evidence or even rumor that he took advantage of his golden opportunity. This does not necessarily mean that he did not, for the Jews, although glad to invent all sorts of stories about "exterminations" to defame the Germans, would not want the world to know that the idea had been put into practice anywhere,[26] while the Chinese would think it inexpedient to shock the tender hearts of the sentimental Aryans. But if Bakony was right in his analysis of the Chinese régime, subsequent events show that Mao missed his chance.

The Polish writer, Louis Bielsky, in another small booklet, *Underground Facts of the Arab-Israel and Moscow-Peking Conflicts*, agreed with Bakony on all essential points and, writing at about the same time, predicted that the Jews would either "promote a revolt against Mao . . . or wait patiently for Mao Tse-tung and Chou En-lai to die, so that the crypto-Jews . . . can

26. The publication of Professor Butz's analysis of the "holocaust" swindle, *The Hoax of the Twentieth Century*, naturally caused among the Jews some dissent as to the policy the race should pursue, and several rabbis, in their own publications and even in their columns in papers for the *goyim*, such as the *Chicago Sun-Times*, issued veiled warnings that too much agitation about their Holohoax might give the Americans ideas they would put into practice. The ruling element in Jewry decided to use their newspapers and boob-tubes to pump a steady stream of sludge in the faces of the dumb brutes; but some intelligent rabbis continued to have misgivings. The *Stratford* (Connecticut) *Express*, 23 September 1978, quoted the opinions of two rabbis concerning the slop currently sprayed from the boob-tubes; one said that the film might make people "wonder why Hitler did not complete the job, and it could encourage Fascism to rear its ugly head again"; the other said that the film was "far too contrived...and could encourage Fascism again." Even the device of ramming the pus into the minds of school children seems to be becoming counter-productive, and one hears that some Jews are coming to feel that they are only advertising their dominion over their American plantation and its livestock. Such dissidents may have been the object of a warning in *Jewish Week*, 29 April 1979: "The Holocaust is our strength. We have been shielded by it for a generation." The blatant hoax, however, may prove to be their great and perhaps fatal weakness.

again get control of Red China . . . and transform it again into a satellite of the Jewish-Soviet Union." One could say that this prediction was verified by subsequent events.

By October 1975, it was obvious that Mao's health was failing and that his old age would soon reach its inevitable end. The Chief Executive of the United States at that time, Rabbi Kissinger, rushed to China for a conference with Mao and found him still so vigorous that the two barely preserved decorum at the farewell dinner, but Kissinger may have scattered in the right places largesse from the pockets of the Jews' beasts of burden in the United States. In the following February, observers were astonished that Hua Kuo-fang had pushed himself into the succession to Chou En-lai and was undermining Teng Hsiao-ping, who was believed to have been in line for the succession to Mao. Richard Nixon, who was then playing the rôle of President in the performances at the White House, was hurriedly sent to China, possibly with another bucket of Americans' money. In April, Hua booted Teng out of the political heaven, and when Mao at last died in September, Hua jumped up onto the vacated throne.

Children and intellectuals often suppose that politicians have principles, and they were astonished when Teng suddenly returned from outer darkness to become second to Hua. Observers surmised, of course, that the two had made a deal for mutual advantage or, to put it less politely, that Teng had obtained his price for selling out his faction. This was verified at once: that faction, now headed by Chiang Ching, who had been Mrs. Mao No. 5 (if I have counted correctly) and was his widow and had the support of three men who had been close to Mao, attempted a revolt to restore "Mao's principles." The revolt was quickly suppressed, and the four leaders were arrested and imprisoned, while the small fry were probably massacred. That was in July 1977, although, oddly enough, Chiang Ching and her confederates were not tried and convicted of "crimes against the state" until January 1981.

Bielsky's prediction seemed to have been verified. The Jews waited for Mao to die and then took over, and what was more, China and the Soviet, as predicted, appeared to kiss and make up their quarrel. Hua and Teng were said to have become real buddies, and the latter negotiated a treaty with Japan, visited the United States, and was rewarded when the United States rushed to him two billion dollars, euphemistically called "credits," although every one knows that the two billion will come from the udders of the American milch-cows. More significant was the errand boy who delivered the cash: he was the Vice President, Mondale, commonly known as "the toast of the homosexuals" and thought by many observers likely to succeed Reagan as the Chief actor in White-House shows, since he can be relied upon to make the white boobs cower before perverts as they now cower before niggers. When the milk of American cows is delivered by so great a man, it must be doubly refreshing.

So far, so good. But then we come to the events of September 1981. Children and intellectuals seem never to understand that friendship between powerful politicians is like the amity that prevails between professional gamblers, who play poker with their derringers ready in their sleeves. Somebody—it was said to be Teng — suddenly drew on Hua, who threw down his cards and vamoosed. And then, after a little interlude with a stand-in, Hu Yao-ping appeared as the dealer, and "old China hands," who know everything, say that he is Teng's man and, despite some official blarney, will deal from Mao's old deck. Chinese names are variously rendered in English, and I do not know whether this Hu is the "Hu Y'o" whom a Chinese defector in 1957 identified as the young man most likely to be Mao's eventual successor, according to Dr. William G. Goddard's *The Story of Chang Lao* (Melbourne, Australia, c. 1962).

If Bielsky and Bakony were right when they wrote, and if the "old China hands" are right about Hu Yao-ping now, the Jews have again lost control of China. And if that is so, it is unlikely that China could be used to attack Japan.

But what if they are wrong? What if Hu Yao-ping holds office by permission of the Jews, whose power must depend on the Yellow Jews in China? I do not know, but I doubt that the Chinese could now be mobilized for an attack on Japan. Some observers feel that the suffering of the Chinese since the Americans delivered them to the Communists has almost completely effaced recollection of the bitterness aroused by the Japanese invasion in the 1930s, and that the constant denunciation of Chiang Kai-shek as a "tool of the white imperialists" has shifted the onus of responsibility to him. And there is a more fundamental consideration. All observers are agreed that the Chinese are now more racially conscious than at any time before in their history, and that the internationalism peddled by Sun Yat-sen and the early Communists is now virtually extinct. And if the Chinese have become so conscious and proud of their

42

race as Mongolians, one wonders whether the Yellow Jews will be able to keep up their masquerade. I do not profess to know, but I suspect that although the Yellow Jews seem indistinguishable from Chinese to your eyes, racially conscious Mongolians will smell out the difference. They may not be as stupid and fatuous as Aryans.[27] And if they are not, the Jews' time in

27. The obtuseness of White men would be incredible, if it were not attested by innumerable examples. The Irish, for example, still venerate the memory of the "great Irish patriot," Robert Briscoe, and his "heroic part in the Irish revolt [against Britain]," his heroism having consisted of inciting murders and planning riots from a place of safety and of smuggling into Ireland arms and bombs that the Irish purchased at high prices from Jewish dealers. They venerate that hero because their newspapers tell them to, and they do so quite oblivious of the fact that "Briscoe" did not have in his veins a drop of Irish blood, being the offspring of Jews who crawled into the island from Lithuania, either before or after his birth. In March 1957, he strutted through Boston at the head of a procession of Irish, suitably adorned while the band played "Wearing of the Green" and he waved his cane at the cheering crowds of "those dumb Micks," as he called them when speaking later to a German-American, although the Jew seems to have concealed his contempt for his dupes when he was with them.

We should not think of such stupidity as a peculiarity of the Irish. In his *Racial Contours* (Douglas, Isle of Man, 1965), H.B. Isherwood, on the basis of his own observations and the latest anthropological data then available, stated that the highest percentage of Nordics was to be found in Sweden, where the Nordic characteristics were more common than in Norway. In my review of Donald Day's book in the January issue of *The Liberty Bell*, I commented on his observation of the Swedes. A reader tells me that he recently attended an academic ceremony at the University of Uppsala: he says the University was swarming with Jews and that the Swedes could not tell the difference between a Jew and a Swede—not because they cowered before the Jewish Terror, which would be understandable, but because they were too stupid to perceive a difference between persons who spoke Swedish. He said that he at last understood that the common phrase, "dumb Swede," did not refer to a person stricken with aphasia or a disease of the vocal chords. The Swedish government has arrested Dietrich Felderer for disrespect to God's Race and has placed him at the mercy of Jewish "psychiatrists." Felderer's crime was to write a book in which he analyzed the hoax called *Anne Frank's Diary*, a piece of fiction so carelessly put together that any person who can read it while awake and fail to recognize it as clumsy fiction is so deficient in common sense that he must be considered intellectually subnormal. It would be bad enough if the Swedish authorities who are persecuting Felderer were doing so in expectation of being rewarded with a few dollars by the Jews, but one cannot exclude the horrible thought that some of them may actually believe the silly story told in the hoax. If they do, they probably believe Grimm's Fairy Tales to be historical records.

China is limited, regardless of whether or not they now control Hu Yao-ping.

I do not pretend to solve the Chinese puzzle. I have tried only to indicate concisely why it is a puzzle.

THE CRUX

So we come back to the question with which we began. And since we are not privileged to receive the revelations that enable so many "conservatives" to know precisely what is going to happen, we have to reason as best we can from the scanty and sometimes ambiguous evidence available to us. That evidence necessarily deals with the present, not some hypothetical future for which we may wish. I do hope most sincerely that our race has latent energies and a will-to-live that is only temporarily in abeyance, but that does not alter the fact that today the United States is not a nation. As Professor Andrew Hacker of Cornell pointed out in *The End of the American Era* twelve years ago, the United States has become a geographical area inhabited by incompatible races and in which our race has been splintered into reciprocally antagonistic groups, each of which tries to profit at the expense of the others. It may be that we can become a nation once more, but the toughest minds today will quail before a reasonable estimate of what that will cost in personal sacrifice, violence, and bloodshed.

In the meantime, Americans who may be distressed by the ever increasing industrial supremacy of Japan can do nothing about it. They cannot try to compete: the gangsters who operate labor unions have us by the throat; in addition, the revolutionary government in Washington has crippled industry and all business by compelling managements to replace Americans with lazy and feckless bums from the sacrosant "minorities" who hate us; and finally our whole economy is being slowly crushed by the ever increasing cost of speeding up the importation and proliferation of every species of our anthropoid parasites. Americans cannot try to destroy Japanese industry by war: even now we are afflicted with mobs of hysterical neurotics who scream with indignation because neutrons cause fission in lithium deuteride. We cannot hide behind tariff barriers: neither the powerful international corporations, once American, nor the gang of counterfeiters called international finance would permit it, and if they were somehow overruled, our worm-eaten economic structure would crumble to dust—and in any case, so long as there are

44

purchasers, they will prefer the superior Japanese product, even at a higher price. We are helpless. We have made ourselves helpless.

The initiative, therefore, lies with the Jews. Will they order action against Japan to keep their American cattle sufficiently prosperous to finance their terrorism throughout the world and also furnish rich nourishment to the hordes that have swarmed into the country that once was ours? I wish I knew the answer.

It may be that we err when we think of the Jewish race as only materialists, predatory for profit. We think of cut-throat methods and dirty tricks to take over the businesses of *goyim* and drive them from the professions, of political corruption and lucrative incitement of depravity. We think of the habitual device used in its simplest form by the parasites who swarmed into the South in the wake of the invading armies in 1865. It was neatly described by Mark Twain. As soon as the ruined plantations were made productive again, a store was promptly established by "a thrifty Israelite, who encourages the thought-less negro and his wife to buy all sorts of things they could do without—buy on credit, at big prices, month after month, credit based on the negro's share of the growing crop; and at the end of the season, the negro's share belongs to the Israelite, and the negro is in debt besides." It is always the same: in Rumania, in Hungary, in Poland, in every country infested by the inter-national race. Of course, the simple method that suffices for Congoids and simple-minded peasants has to be made more elaborate and sophisticated when applied to prosperous Aryans, including millionaires, but the principle remains the same. Where there is blood to be sucked from the natives, the leeches are always fat.

That picture of the invading hordes is accurate so far as it goes, but it may not be complete. A race is more than an aggregation of individuals, and it is as much a spiritual as a physical entity. And there are historical incidents in which the spiritual force of Judaism has unmistakably overcome the selfish interests of individuals. A famous example is an event in A.D. 117, which was summarized in Ralph Perier's little booklet, *The Jews Love Christianity:*

> In the capital city [Cyrene] of that prosperous province [Cyrenaica] of the Roman Empire, the Jews, naturally, had planted a huge ghetto and they undoubtedly controlled a large part of the trade on which the province's prosperity

depended. Many Jews must have been among the wealthiest inhabitants. But, nevertheless, the race's innate nihilism was excited by a christ, who announced the glad tidings that Yahweh had said that the time had come to put the *goyim* in their place. Filled with a zeal for righteousness, the Jewish swarm caught the stupidly complacent Greeks and Romans off their guard and slaughtered more than 200,000 men and women in various ingenious ways .... God's People then destroyed all the property in the city (*including their own!*), apparently by burning the city and then levelling to the ground such walls as remained standing. They then rushed out into the countryside to destroy the villages and uproot the crops.

I italicized the significant phrase. We do not know whether the wealthy Jews whose riches were thus annihilated were overmastered by the mob or had themselves caught the enthusiasm for ripping the guts out of people whose civilization and culture the race has always hated, but what is more significant is that *every* one of the Jewish rabble, even the very poorest, must have had to abandon and sacrifice his possessions, however few they may have been, when he and his fellows were inflamed by a spiritual ardor.

A race's innate character is most clearly shown in its favorite myths. No reader of the "Old Testament" can have failed to notice that while there are many tales of highly profitable theft, subversion, and looting by God's Chosen Bandits, the real *Leitmotiv* of the whole collection is destruction, universal massacres and total destruction. There is the well-known passage (*Exod.* 23.27) in which Yahweh promises each and every Jew that he "will destroy all the people to whom thou shalt come."[28] And the story narrates, over and over again,

---

28. The words that I have quoted from the King James Version are attenuated in later translations on the basis of quibbles about the Hebrew text that are not worth mentioning. Very significant, however, is the meaning of the Hebrew text that was current in the first century B.C. and was translated into Greek in the Septuagint. In it the crucial words are ἐκστήσω πάντα τὰ ἔθνη, and thus Yahweh promises "I will befuddle the minds of all the gentiles [=goyim]." This text agrees with the rest of the chapter, in which Yahweh explains that the hated races are not to be exterminated all at once, but gradually and "little by little." That, in turn, fully agrees with the explanation given by Philo Judaeus, the Jews' great apologist of the First Century (A.D.). Admitting that the tales about the conquest of Canaan

with wearisome iteration, the triumphs of the blood-thirsty and nihilistic marauders. Yahweh's special pet, Moses, boasts, "And we took all the cities [of Bashan] ... threescore cities ... and we utterly destroyed them ... utterly destroying the men, women, and children of every city." And although Moses didn't get to enjoy much more carnage, the savage swarm moved on to Jericho, "And they utterly destroyed all that was in that city, both man and woman, young and old, and ox and sheep and ass, with the edge of the sword . . . . And they burnt the city with fire, and all that was therein." "And Joshua[29] ... utterly

---

were intrinsically unbelievable, he gave a rational explanation of them (*Hypoth.* 6.6-7 = 356d-357a). When the wandering tribe of Jews reached Canaan, intending to slaughter the natives and take their country from them, the Jews were necessarily incapable of armed aggression against a strong nation, but the Canaanites were so befuddled that they believed their implacable enemies to be a godly and peaceful folk and accordingly invited the Jews into their country and permitted them to set up their synagogues and colonies. That proves that the Jews are God's People, because God must have made the Canaanites so stupid as to let the Jews immigrate. Of course, when the Jews had securely lodged themselves in the country they intended to steal, they destroyed the gullible *goyim* by methods, doubtless including their habitual technique of subversion and inciting civil discord and war, that Philo thought it would be tactless to describe. We may be virtually certain, therefore, that the Septuagint preserves the meaning of the original text, although later tales in the collection lovingly describe a military invasion of Canaan and the delights of slaughtering its inhabitants. One Jewish hoax that long imposed on our people was the claim that they sedulously preserved the texts of their holy books without alteration; that was, of course, definitively exposed by the few Dead Sea Scrolls that have been published, and is probably one reason why the Jews, with, of course, the complicity of the Christians, have made certain that the great bulk of those scrolls will never be read by honest *goyim*. (The story now is that a mysterious infection has attacked the organic fibers of the scrolls and is turning them into gelatin, so they are now said to be locked up in lightless vaults and one of the "custodians" has boasted that no one will ever again see them.)

29. The King James Version (and, so far as I have noticed, all others in English) is in error here. The name of the supposed leader of the Jewish invasion and despoliation of Canaan should be spelled 'Jesus' since that is the spelling of the same name when it refers to the protagonist of the "New Testament." The name is the Hebrew YŠW, and since vowels were not written in Hebrew, it was easy to deceive persons who did not know Hebrew or the language from which that dialect was derived by supplying different vowels in the two contexts. In the last centuries B.C. and early centuries A.D. the name was pronounced as *Yēshua* or *Yēshwa*, which, as filtered through Greek and Latin, gives the English 'Jesus.' Of this, there

destroyed all the inhabitants of Ai ... And Joshua burnt Ai, and made it a heap for ever, even a desolation unto this day." And so the inspiring tale goes on and on and on. "So Joshua smote all the country of the hills, and of the south, and of the vales, and of the springs, and all their kings: he left none remaining, but utterly destroyed all that breathed." A righteous lust to kill all the men, all the women, all the children, all the animals, everything that breathed, and to destroy cities and make of them mounds of desolation in a desert, was stronger even than the greed of the godly brigands whose piety is celebrated in their exemplary tales.

Their "prophets" attain a memorable eloquence when they are inspired by visions of world-wide death and desolation. "The indignation of the Lord is upon all nations; ... he hath utterly destroyed them, he hath delivered them to the slaughter. Their slain shall be cast out, and their stink shall come out of their carcasses, and the mountains shall be melted [!] with their blood. And all the host of heaven [i.e., the constellations] shall be dissolved, and the heavens shall be rolled together as a scroll."—"Their land shall be soaked with blood, and their dust made fat with fatness [of decaying flesh] .... And the streams thereof shall be turned into pitch, and the dust thereof into brimstone, and the land thereof shall become burning pitch. It shall not be quenched night or day; the smoke thereof shall go up forever: from generation to generation it shall lie waste; none shall pass through it for ever and ever."—"I will break in pieces the horse and his rider ... I will break in pieces the chariot and its rider ... I will break in pieces man and woman ... I will break in pieces old and young ... I will break in pieces the young man and the maid ... I will break in pieces the shepherd and his flock ... I will break in pieces the husbandman and his yoke of oxen ... I will break in pieces captains and rulers ... And the land shall tremble and sorrow: for every purpose of the Lord shall be performed ... to make the land of Babylon a desolation without an inhabitant."—"I [Yahweh] have cut off the nations: their towers are desolate; I have made their streets waste, that none passeth by; their cities

---

can be no possible doubt: in the Septuagint the character whom ignorant Christians call 'Joshua' is uniformly called 'Jesus,' and the book that is called 'Joshua' in the Christians' Bible is entitled 'Jesus.' The fact that there was only the one name is admitted by Christian theologians, but they maintain the false distinction for business reasons.

are destroyed, so that there is no man, that there is no inhabitant ... My determination is to gather the nations, that I may assemble the kingdoms, to pour upon them mine indignation, even all my fierce anger: for all the earth shall be destroyed with the fire of my jealousy."

The same spiritual *Leitmotiv* of Judaism appears vividly in the apocalypse that was selected for inclusion in the "New Testament." It would take too long to enumerate the ingenious ways in which Jesus afflicts, tortures, and kills all the inhabitants of the earth, and every reader of that mad phantasmagoria will remember that Jesus, in a paroxysm of nihilistic fury, destroys the mountains and the seas, destroys the whole earth, destroys the sun and the moon, destroys all the stars—destroys the whole universe, destroys everything, destroys and destroys.

No other mythology so reeks of an insane lust to torture, to kill, to destroy, to create only desolation and nothingness. And this spiritual force has characterized the activities of the Jews throughout history: they can only destroy. And a few Jews to whom we should be profoundly grateful, notably Marcus Eli Ravage, Oscar Levy, and Maurice Samuel, have been so candid as to tell us the truth explicitly: "We are intruders, we are subverters."—"We Jews ... are today nothing else but the world's seducers, its destroyers, its incendiaries, its executioners."—"We Jews, we the destroyers, will remain destroyers forever."[30]

This is a cardinal fact that we must take into account in our estimates of the present. It is obvious that the Jews derive great profits from many forms of subversion—from pornography and the incitement of degeneracy, from class warfare, from wars between nations of our race, from the inflation of counterfeit currencies and the impoverishment of our people, and from many similar activities.[31] But if we consider such things from

30. The passages from which I have taken these sentences are more fully quoted, with bibliographic references, by Colonel Farrell in his article in the March issue of *The Liberty Bell*, p.31.

31. This includes, of course, the instigation of destructive lusts in the natives. Malcolm Muggeridge, writing in *Time*, 3 December 1979, proposed an explanation of the intensive campaign in our schools and newspapers to spread the race-destroying plague of homosexuality. His explanation

the standpoint of the race, not from the standpoint of the individual Jew who battens on us, is it not likely that the material profit counts for much less than the spiritual satisfaction? And if we consider some of the Jews' work, I cannot see how it could conceivably yield a net profit. What monetary gain can they have obtained, or intended to obtain, by spending vast sums to incite the niggers to rape, murder, and arson? What profit from destroying civilization in Rhodesia and making that land again a land of savages? What can the Jews in South Africa gain in material terms from their present intensive effort to destroy the white population and make of that country another Rhodesia? Is it not obvious that they could squeeze much more money out of the White population by peaceful parasitism and without inciting the racial hatreds that disrupt the economy and could conceivably bring retribution upon themselves? The only explanation, it seems to me, is that with their race as a whole spiritual considerations are paramount, paramount over profit and even over self-preservation. One can foresee the logical end in a future that may not be too distant: one can see the last Jews dying with exultation on the surface of a planet from which they have exterminated all other human beings, all animals, all vegetation, all life—a planet of which they have made "a desolation of desolations."

If this analysis of the Jews' racial instinct is correct, it answers our initial question. It is most unlikely that the Jews will wish to abate the growing industrial supremacy of Japan so long as its effect is to weaken us, induce economic prostration, and accelerate our race's already vertiginous progress to extinction.

---

merits consideration. He believes that John Maynard Keynes, for example, incensed by the loss of the ministrations of a favorite pervert, took vengeance on society "by inventing an economic theory which, after a period of spurious prosperity, must infallibly bankrupt the countries which adopt it." The article is accompanied by a photograph which reminds us that Keynes can have been only partly an Englishman; I do not know whether or not the non-Aryan race that entered into his composition was Jewish, but it is well known that intelligent mongrels usually feel a bitter rancor against the society that made them possible. Muggeridge by implication also accuses E.M. Forster and Lytton Strachey of the same social incendiarism, but, so far as I know, they and the other noted homosexuals whom he mentions were of uncontaminated English descent.

www.ingramcontent.com/pod-product-compliance
Lightning Source LLC
Chambersburg PA
CBHW032019190326
41520CB00007B/549